Look-Alike Animals

# IS IT A TURTLE OR A TORTOISE?

by Anita Nahta Amin

PEBBLE
a capstone imprint

Pebble Sprout is published by Pebble, an imprint of Capstone.
1710 Roe Crest Drive
North Mankato, Minnesota 56003
www.capstonepub.com



**Library of Congress Cataloging-in-Publication Data**
Names: Amin, Anita Nahta, author.
Title: Is it a turtle or a tortoise? / by Anita Nahta Amin.
Description: North Mankato, Minnesota : Pebble, [2022] | Series: Look-alike animals | Audience: Ages 5-8 | Audience: Grades 2-3 | Summary: "An animal that looks like a turtle munches on some grass. It has scales and a hard shell. But is it a turtle or a tortoise? Turtles and tortoises are similar, but they have some important differences. Find out how their legs, feet, shells, and behaviors can all help you tell these two animal look-alikes apart. Filled with stunning photos and playful text, early learners will be delighted as each page turns"— Provided by publisher. Identifiers: LCCN 2021004160 (print) | LCCN 2021004161 (ebook) | ISBN 9781663908674 (hardcover) | ISBN 9781663908643 (pdf) | ISBN 9781663908667 (kindle edition) Subjects: LCSH: Turtles—Juvenile literature. | Testudinidae—Juvenile literature. Classification: LCC QL666.C5 A45 2022 (print) | LCC QL666.C5 (ebook) | DDC 597.92—dc23 LC record available at https://lccn.loc.gov/2021004160
LC ebook record available at https://lccn.loc.gov/2021004

**Image Credits**

Alamy: Scubazoo, 23; Getty Images: passion4nature, 19; Newscom: Photoshot/NHPA/Joe Blossom, 18; Shutterstock: Andrzej Grzegorczyk, cover (bottom), Anna_Kova (design element), cover (middle) and throughout, Bildagentur Zoonar GmbHm, 28, blue-sea.cz, cover (top), Bluesnaps, 30, Breck P. Kent, 31, Carolyn Smith1, 27, Chai Seamaker, 29, damann, 4–5 (bottom), EcoPrint, 10, 25, Fotogrin, 15, Gerry Bishop, 26, HeavilyMeditated, 20, John Krampl, 21, Jurgens Potgieter, 22, Kirill Skorobogatko, 5 (top), Kjeld Friis, 14 (top), Mauro Rodrigues, 8, meunierd, 16–17, Michael Woodruff, 7, Muan Sibero, 24, Ondrej Prosicky, 4 (top), seasoning_17, 3, 9, Tukaram.Karve, 12–13, vitormarigo, 6, Vlad61, 11, Vladimir Wrangel, 14 (bottom)

**Editorial Credits**

Editor: Carrie Sheely; Designer: Elyse White; Media Researcher: Svetlana Zhurkin; Production Specialist: Laura Manthe

Printed and bound in China. 5241

A desert animal eats a cactus. The animal has a high, round shell. It has thick legs. Is this animal a **turtle** or a **tortoise?**

A tortoise is a kind of turtle. But it's **different** from other turtles. Let's find out how to tell them apart!

Shells aren't just at the beach! Turtles and tortoises have **shells** around their bodies.

Most shells are **hard** and **rounded.** But some turtles have **soft** shells that are very **flat.**

Most tortoise shells look like **upside-down bowls.** They're **high, round,** and **heavy.** Most turtle shells are **flatter** and **lightweight.**

You can **swim** with turtles but not tortoises. Tortoises live on **land.**

yellow-footed tortoise

snapping turtle

Most turtles live in or near **water.** Some, like **sea turtles,** live in the **ocean.** Others, like **snapping turtles,** live in **ponds** and other **freshwater.** A few, like **box turtles,** live on **land.**

## ONE SHELL FOREVER

You can change your clothes, but tortoises and turtles **can't change their shells**. The **shell** doesn't come off. It's part of the body. The animal can feel when its **shell** is being touched.

The **shell stays** with the turtle or tortoise

**forever.**

As the animal grows, the **shell grows** too!

mother tortoise and her baby

# Danger is near!

Tortoises pull their heads, tails, and legs **into their shells.** So do many turtles. This keeps them **safe** from animals that are trying to attack them.

Some turtles, like sea turtles, can't tuck body parts into their shells. They **speed away** instead.

sea turtle

# How do you get off your back?

Your arms and legs can help. Similarly, some turtles and tortoises can use their **heads, legs,** and **feet** to **flip over.**

Other tortoises **rock their legs.**

**One, two, three,**

**flip!**

# Glide!

Sea turtles have **flippers** to help them **swim.** Other turtles that spend a lot of time in water have **webbed feet** for swimming.

Tortoises **can't swim.** They have **thick legs** for walking on land. Their hind legs are shaped like **tubes**. Their toes have **claws.** They can use their claws for **digging.**

Tortoises
and turtles
breathe
air.

Turtles can hold their breath **underwater.** Some hold it for **hours.** They come up to the water's surface to breathe. Tortoises never need to hold their breath because they live on land.

turtle

# Brr!

## It's freezing!

Until it warms up, both animals can enter a **resting state.** Their body systems slow down. They might rest for **months.**

Tortoises rest in underground tunnels called **burrows. Land turtles** snuggle **under soil** or **leaves.** Other turtles go **under rocks** or in **mud** at the water's bottom.

tortoise

Your body temperature doesn't change on a hot day. But turtles and tortoises are **reptiles.**

**Reptiles are cold-blooded.**

Their **body temperature** changes with the temperature around them.

When they're **too hot,** they might go **into the shade** to stay cool. They also might stay in their burrows or sit in cool water.

To **warm up,** they **sunbathe.** Sea turtles swim to warmer water.

turtle

Which reptile is faster?

Turtles would win a race by a lot!

The fastest tortoise ever recorded walked 0.92 feet (0.28 meters) per second.

leopard tortoise

Leatherback sea turtles can swim as fast as **22 miles (35 kilometers) per hour.** This is more than 32 times quicker than the fastest tortoise!

Can you remember the way to places you visited a long time ago?

# Tortoises and turtles can!

sea turtle laying eggs

**Sea turtles lay eggs** where they were born. **Tortoises dig wells** to catch rainwater to drink. Later, they **remember** where the wells are.

TALK IT OUT

# Listen up!

Both animals make sounds. Turtles **hoot, grunt,** and **chirp.** They might **hiss** when **scared.**

turtle

Tortoises **cluck** and **whine.**
If they **bob** their heads,
**stay away!**
They could be **mad.**

# It's lunchtime!

Almost all tortoises **eat only plants.** They eat grass, bushes, and flowers. Most turtles **eat** both **plants** and **meat.**

tortoise

# Neither reptile has teeth.

They use their sharp beaks to rip apart food. Some turtles and tortoises have beaks with jagged edges. They can crush food inside their mouths.

turtle

# Tortoises can live longer than turtles.

Most **tortoises** live **50** to **100 years.** When cared for by people, some can live nearly **200 years!**

# Sea turtles can live past the age of 60.

Other turtles usually live for 20 to 30 years.

# IS IT A TURTLE OR A TORTOISE?

1. A reptile swims in the ocean its whole life. Is it a turtle or a tortoise?

2. An animal feasts on jellyfish for dinner. Is it a turtle or a tortoise?

3. A reptile uses its thick legs to dig a tunnel. Is it a turtle or a tortoise?

4. It buries itself in mud at the bottom of a pond during winter. Is it a turtle or a tortoise?

5. An animal with a high, round shell turns over by rocking its legs back and forth. Is it a turtle or a tortoise?

Answer Key: 1. turtle 2. turtle 3. tortoise 4. turtle 5. tortoise